My dog thinks he's human!?

AF719604

First published in Great Britain
in 2009 by Spruce, an imprint of
Octopus Publishing Group Ltd
Carmelite House
50 Victoria Embankment
London EC4Y 0DZ
www.octopusbooks.co.uk

An Hachette UK Company
www.hachette.co.uk

The authorized representative
in the EEA is Hachette
Ireland, 8 Castlecourt Centre,
Dublin 15, D15 XTP3, Ireland
(email: info@hbgi.ie)

This edition published
in 2026 by Hamlyn

Text copyright © Peter
Wedderburn 2009, 2026

Illustration copyright ©
Per José Karlén 2009, 2026

Distributed in the US by
Hachette Book Group
1290 Avenue of the Americas,
4th and 5th Floors
New York, NY 10104

Distributed in Canada by
Canadian Manda Group
664 Annette St., Toronto,
Ontario, Canada M6S 2C8

All rights reserved. No part of
this work may be reproduced
or utilized in any form or by any
means, electronic or mechanical,
including photocopying,
recording or by any information
storage and retrieval system,
without the prior written
permission of the publisher.

Peter Wedderburn asserts the
moral right to be identified
as the author of this work.

ISBN: 978-0-60064-006-6
eISBN: 978-0-60064-007-3

A CIP catalogue record for
this book is available from
the British Library.

Printed and bound in China.

10 9 8 7 6 5 4 3 2 1

Publisher: Lucy Pessell
Senior Editor: Tim Leng
Designer: Isobel Platt
Assistant Editor: Samina Rahman
Production Manager:
Caroline Alberti

FSC www.fsc.org MIX Paper | Supporting responsible forestry FSC® C144853

Peter Wedderburn
Illustrated by Per José Karlén

My dog thinks he's human!?

Practical advice for dog lovers

hamlyn

Contents

Introduction

Dogs share our homes and our lives in a closer way than any other pet. A dog becomes as much a member of the family as any member of the household, included in photographs and storytelling. Humans often treat their dogs like little humans, and so it's no wonder that they behave as if they are one. But dogs are just that...dogs! And to be happy pets, they need to be treated like dogs, so that they think they are dogs.

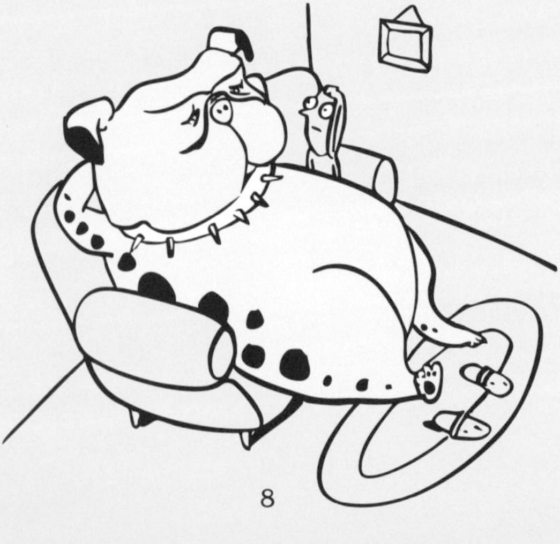

This doesn't mean you need to be a tough disciplinarian at all times, nor do you need to whisper quiet messages into your dog's ears. But you do need to be Top Dog, and you need to discover that sometimes the best way to treat your dog is different to what you may feel like doing. There are times for spoiling dogs and times for being more assertive. Every dog is different and while there are no universal rules, it is possible to have broad guidelines that apply to every dog and owner. That's what this book is about. It includes 50 suggestions for looking after your dog in the best possible way, including tips on health and training.

Life together should be a pleasure – for human and dog. The aim of this book is to help you both enjoy your relationship to the full.

Chapter 1

A dog's dinner

How can I be sure that their diet is balanced?

Their choices are:

1. Moist dog food with dog 'mixer' biscuits.

2. Dried 'complete' food.

3. Home-cooked meals.

Your vet will tell you what they think is good for your dog and Fido will tell you what they think is tasty! If you put down what's good for them and they eat it, they'll be happy and healthy. If they choose not to eat it, tweak the ingredients a bit, but stay this side of the 'what's-good-for-them' fence and soon you'll strike the right balance. This may be a mix of moist food and mixer biscuits or a dried 'complete' food designed specially for dogs, but don't forget the fresh water beside their food bowl. Don't stuff them on cakes, curries and spaghetti Bolognese, and do remember that Fido, like you, has discerning taste buds. It's OK to spoil them from time to time. Occasional portions of delicately seasoned fish and vegetables or a slice of roast beef with gravy will be enthusiastically received.

Posh nosh or basic rations?

Commercial dog food varies as much in quality as it does in price. You can take the nutritionist's word on this or you can try a simple experiment at home for yourself. Feed your dog a high-quality diet for two months and judge the difference for yourself.

Rule of thumb: dogs that are fed on quality foods tend to look bright-eyed and bushy-tailed, while those fed on a cheaper diet, which consists of a high percentage of indigestible bulking agents, not only look less healthy but also require a correspondingly higher percentage of poop-scooping!

When it comes to dining, watch out for the small print.

Your dog is what you feed them, so if they're hyperactive and uncontrollable, check their diet. Avoid the foods that have a bewildering list of additives and derivatives, and choose brands with ingredients that you can understand. You wouldn't give a child a handful of candy containing tartrazine after it has been proven that it causes hyperactivity in children. Your pooch is just as sensitive.

Delicious treats or toxic truffles?

The stimulants in chocolate are highly poisonous and potentially fatal to dogs, causing their heart to beat too rapidly. The toxic dose depends on the weight of the dog, so that smaller dogs are most vulnerable. Whilst dark chocolate is better for us, it is three times more toxic than milk chocolate for dogs, and as little as 25g (1oz) could be enough to kill a 3kg (6½-lb) Yorkshire terrier. Store all chocolate out of dog-reach – most fatalities happen when dogs consume entire boxes of chocolates that have been accidentally left within grabbing distance.

You may feel guilty looking into those big sad eyes, but you may just be saving Fido's life by saying No. Find another type of treat that is safe for them. Chopped-up cooked sausages or dried liver treats may not appeal to you, but your dog will adore them.

A butchers' bone – can it harm them?

Yes. The sharp pieces of a raw bone can get lodged in their digestive tract, causing an obstruction that requires complex surgery to resolve. In addition, the small fragments of bone can accumulate and solidify in the lower bowel, forming a concrete-like block that can be very difficult to remove. It's safer to give them manufactured nutritional chews, or a chew toy such as a Kong, stuffed with food. If you really want to give them bones, buy some pre-cooked beef shins and knuckles from the pet shop. These are less likely to splinter, and won't carry the risk of a bacterial infection that could be passed on from raw bones.

Does eating grass mean they're vegetarian?

Dogs are omnivorous, enjoying both meat and vegetable components in their diet. And most do enjoy occasional grazing on grass, picking out leafy stems and chewing contentedly. Nobody knows exactly why dogs do this; perhaps for additional nutrients that may be missing from their diet? Some dogs vomit after eating grass, suggesting that they deliberately make themselves sick if they are feeling a little nauseous. There is probably no single answer, and every dog may chew grass for a different reason. It's nothing to worry about, as long as the grass is free of pesticides and it causes no other physical problems.

My dog has a gassy problem. What can I do about it?

Nobody enjoys the company of a flatulent dog. If pooch has a gassy problem, look at their protein consumption. Introduce them to a specifically designed 'highly digestible', dry biscuit diet, available from your vet or pet shop. Feed them smaller portions more often. Alternatively, try a different meat source, such as duck, fish or venison sold as special single-protein diets available through vets. If these changes don't clear the air, then either visit your vet or invest in some gas masks.

The vet says my dog is fat, but I think they look cute. Who's right?

If the vet says your dog is overweight then they ARE, no matter how hard you try to convince yourself that they are merely 'well covered', and at five years old, it isn't puppy fat any longer. The loving solution is to put porky pooch on a diet and not be tempted to give unnecessary treats. Research some light diets online to find one that has the right nutritional balance for your dog, then stick to it. The most effective low-calorie diets are kept as 'prescription-only' products, supplied by vets for those patients that really need them. Many vets have free obesity clinics with regular weigh-ins so that you can track pooch's progress back to healthy happiness.

My dog barks at me for a cup of tea. Is it OK to give them some?

No! Caffeine in tea and coffee can be toxic to dogs. A 284ml (half-pint) cup of tea contains around 60mg of caffeine, instant coffee up to 100mg, and the real stuff a colossal 175mg. A skinny latte may be fairly safe for a large Labrador, but half a cup of tea could be enough to cause serious problems for a toy poodle. They may bark for a cuppa, but if you ignore their commands, they'll soon get the message.

Chapter 2

From mangy mutt to pampered pooch

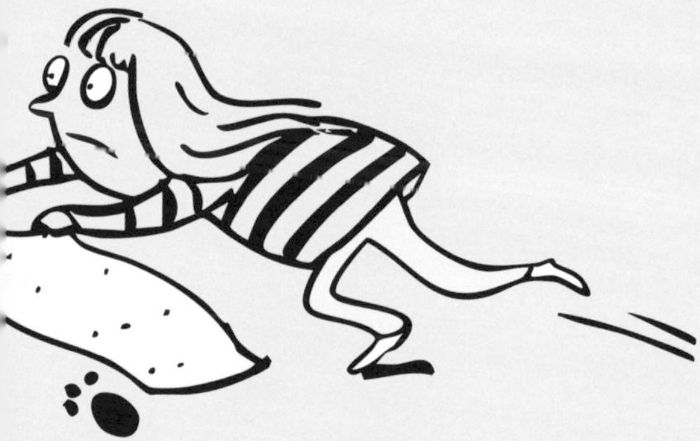

How can I make sure their friends don't share their fleas with them, too?

The long and the short of it is that fleas jump from dog to dog regardless of who owns them. Whether your corgi is of royal stock or a tearaway, you should use a once-monthly vial of flea drops, which is applied to the skin between your dog's shoulder blades. This leaves a continual presence of anti-flea substance in his skin or bloodstream, so that if a flea lands on him, it will not survive for long. Ask your vet about the best version of drops available and suggested usage. If your dog does get fleas, you will also need to treat for tapeworms, which are transmitted by dogs swallowing the fleas that they nibble out of their own coats. And remember, the rich itch too!

Don't let your dog give grooming the brush off.

From a dog's perspective, it's easy to see why grooming may look like a game. You come towards them, holding a 'toy' in your hand (they don't know it's a brush). You wave it around, and when they try to grab it, you wave it even more. To them, it looks like the stuff you do in the park. You need to combine basic dog training with gentle grooming. Teach them to sit and stay on command, and keep rewarding them (with praise and treats) as long as they remain sitting still. When they can do this, take out the brush, and do some simple grooming, continuing to reward them for behaving well. Do a little bit each day and in no time scruffy pooch will be looking sleek.

They drench the house after a bath. Will a bathrobe help?

Luxury warm and snug bathrobes are available for dogs but manufacturers warn that they are a fashion accessories and should be worn only under supervision. A better way to avoid saturating the house is to plan 'bath day' more carefully. The options are:

1. On a fine day, wash outside, where they can race around drying *au naturel.*

2. Limit the affair to just one room and line it with newspaper.

3. Place an extra-large, high-absorption towel in their bed. Towel dry your dog as best you can, then entice them into bed, with some attention-grabbing device to keep them occupied.

4. Give up. Use a professional groomer.

My dog is moulting. Are they likely to go bald?

You can put away that hair-restorer because it's normal for dogs to shed fur in small quantities all year round. Once or twice a year also, most dogs have a heavy moult. In the spring, the long, dense winter coat is replaced by a lighter summer coat, with the process happening again in reverse in the autumn. Despite this normal hair loss, it's unlikely your dog will ever have bald patches.

Frequent grooming is the best way to prevent moulting hair getting onto your carpets and furniture. Other ways to remove hair include taking your dog for walks in dense undergrowth and giving them a bath and towelling them dry thoroughly. And if you have an exceptionally laid-back pooch, try the vacuum cleaner.

Why does my dog refuse to wear the lovely jacket I bought them?

Do you blame them? Dogs only wear clothes because their owners make them. Forcing a dog to wear clothes when they don't want to is tantamount to cruelty.

That said, there are several scenarios when clothing for dogs can be useful: for example, if a long-haired dog has been clipped short, they can get cold and might appreciate an extra layer. Fine-haired or very small dogs might be glad of protection from heavy rain, a dog with an allergy to grass may enjoy a protective layer and a dog who for medical reasons has lost large patches of their coat can be misconceived as having been mistreated by its owner, so a light cover might avoid confused protests and embarrassments in the street. Otherwise, go with what nature intended, and if anyone must be forced to wear a tartan vest and booties with pom-poms, better that it is you!

It is nail-biting stuff! Why won't they let me clip their nails?

The closest thing I have encountered to a Great White Shark is a Welsh terrier who didn't like having their nails trimmed! Dogs are drama queens and just the mention of nails can send them into absolute fits of distress, howling and growling. And rightly so: if you cut too much nail, you could end up nicking the quick, causing pain and bleeding.

Start gradually when they are still a puppy and take the process a bit further each time, first by physically restraining them and rewarding them as long as they stay calm. Next time, pick up one foot, and examine it, then each foot in detail. If they let you do this without a fuss, move on to clipping the sharp tip of one nail. Remember to reward and praise all along. Do two nails the following day and so forth. If they start to struggle, go back a step. Any lack of confidence on your part is quickly telegraphed to your dog, and remember, a dog has teeth!

Silver threads amongst the gold. Can you get hair dye for dogs?

Here we go again! Can't a dog age in peace without human vanity getting in the way? For heavens sake, dogs age too and a salt-and-pepper coat can look distinguished. Besides, one should be asking what can be done to extend their life and make their old age comfortable, not how to cover it up. Many older dogs begin to suffer from perfectly treatable problems that are not recognized by their owners, including arthritis and dental disease. Speak to your vet, who may suggest a special diet designed for older dogs that may help to optimize their physical state, mental health and appearance in their later years, and save the blue rinse for yourself.

My dog's teeth have gone brown. Should I start brushing them?

To optimize dental health, all dogs should have their teeth brushed on a regular basis from puppyhood onwards. The question is this: do you really want to get into your dog's mouth? Most people accept the trade-off: they don't brush their dog's teeth, and their pets develop moderate dental disease as they get older. If your pet has got to the stage where they have significant build-up of tartar, you will need to ask the vet to clean up their teeth under anaesthesia. If you do choose to brush their teeth and give them that Hollywood smile, ask your vet about the best toothbrush and toothpaste to use. Anyone for raw chicken-seasoned toothpaste?

Chapter 3

Canine crimes

How can I teach them not to bite when we're playing?

Puppies learn 'bite inhibition' between the ages of 2 and 12 weeks from their siblings, but since they go to new homes at around 8 weeks, it is up to new owners to complete the life lessons about how hard to bite without causing harm. No, I am not asking you to bite your puppy. But if they bite you, you should say 'ouch' loudly, and withdraw your attention. If they do it again, say it more loudly, and leave the room until they learn that certain biting results in adverse consequences and stops doing it. As they grow, teach them in a similar fashion not to grab people, however gently. Provided you remain in control, there is nothing wrong with a little rough and tumble, but if things do get out of control, step back and consider whether you are teaching them the right lessons. If not, get help from a professional dog trainer to put their development back on track.

How can I get my dog to tone down their hysteria when I get home?

This is a common problem and easy to fix. You have accidentally trained your dog to be like this. When you come home, and they give you lots of adoring attention, you respond by greeting them enthusiastically in return, rewarding them for behaving in that way.

To re-train them, you have to behave in a way that seems intuitively wrong if not cruel: you ignore them completely when you get home, and act as if they aren't there. Just do your own business while blanking them completely. When they don't get a response from you, they will eventually settle down. Once they have calmed down, you can then call them over to you, and greet them as they have never been greeted before. If you adhere to this new routine for a few weeks, you will find that they become a much-composed beastie.

To pee or not to pee?

None of us likes to come downstairs half asleep, barefooted, only to be shocked into the new day by stepping in something wet and stinky on the floor. Shocked as you may be, rubbing your dog's nose in their mishaps or shouting at them is counterproductive. It takes about 14 weeks for a puppy to be house trained, so until then use your own radar to seek out the tell-tale signs of when they are likely to need to go. When they wake, soon after eating, when they seem restless or begin that peculiar, directionless circling walk dogs do with head down and tail up, get them outdoors. When they do get it right, make a fuss and reward them. You'll soon be back to being barefoot with confidence.

When is a bed not a bed?

So, your dog has taken over your sofa! Well what are you going to do about it? Be a pussycat and let them take control? Get back to being leader of the pack, reclaim what is rightfully yours and show them where their own bed is. From pooch's perspective it is probably down to you that they are there in the first place. Something you have done has made them feel it is OK for them to be there. Now you have to break the habit and create a new rule...and stick to it. Make the sofa less appealing to them than their own bed...place some treats in their bed, reward them when they go and settle there and perhaps plug-in a DAP diffuser to spray soothing pheromones into the air beside them. Soon they will be comfortable knowing their place and you can be king of the castle again.

What can I do to ensure my new dog doesn't turn out neurotic?

In the first place, choose a dog with the genes of a calm dog! If you are getting a puppy, try to arrange to meet both parents to see if they are obedient, relaxed dogs. If they are, then there is a strong possibility your puppy will turn out the same way. If they are bloodthirsty, salivating beasts with a killer glint in their eye and a growl to match, avoid their puppy for the same reasons!

Don't select the nervous shivering mutt in the corner just because their big, sad eyes make you want to hug them! After all, what you are looking for is an emotionally stable, cheerful and contented dog. Any signs to the contrary should be taken as warning bells unless you are a skilled dog trainer or canine psychiatrist.

Are you, as their owner, as well balanced as the dog you seek? Or are you a little over-anxious, over-sensitive and obsessive? With regular obedience training, socializing, routine walks and, if necessary, a good dog trainer, owners and their dogs become as well adjusted and well behaved as each other.

Should I be feeding my dog from my plate?

When it comes to mealtimes it pays to remember just who is Top Dog! Feed your dog from your plate and soon they'll expect you to set a place at the dinner table...and the breakfast bar too! Your dog will gladly chow down on anything they're offered, so it is your responsibility to give them the balanced diet they need to stay healthy, otherwise the person to blame when they've grown too fat to groom their extremities will be YOU! Weigh them regularly and use common sense. If you already have a chubby salivating canine hovering by the table, place them in another room at mealtimes.

Is it true that there are high chairs for dogs?

It is possible to buy a high chair specially designed for dogs, complete with a lead clip so they don't irresponsibly clamber out of the chair and all over your elegant table arrangement. But it is probably smarter to recognize that they are a dog and happy dogs should learn to sit quietly wherever is appropriate during dinner. They may appear to thrive on pampering but it is vital to have clear lines that separate the species. A dog sitting at the table for dinner has clearly strayed into human territory.

Why does my dog sledge their bottom along the ground?

Suddenly your dog seems to have learned a strange new trick. They place their bottom on the floor and drag themselves along as though riding on a sledge. It is possible that they have worked out that this activity gets extra attention from you, even if that attention involves you being annoyed with them. The message is simple: don't let them annoy you, because this will only make the problem worse.

The bottom sledging is not a new trick at all, but probably one of three things. Firstly, anal glands can become clogged and sometimes infected: your vet may need to attend to these. Secondly, some worms can cause itchiness beneath the tail, so give a broad-spectrum worm dose at least once every three months. And finally, there is a long list of skin diseases that can cause localized itchiness: again, this is one to work through with your vet. Pooch doesn't mean to entertain you or to annoy you... they've just got an itch!

Why does my dog do what they want, not what I tell them to do?

It's no good trying to reason with your pet as though they were human. Dogs need to be told and to learn to do what they are told. A successful owner–dog relationship depends upon you being in control and teaching him appropriate boundaries. If a dog does what they want in small, harmless ways, it is likely that they may start to do what they want in big, annoying, even dangerous ways. Before you know it, they will growl when they don't get their way, possibly biting people who try to interfere. If you are not confident about retraining your canine lout into a happy socialized dog, seek professional help. Otherwise, prepare to get snarled at!

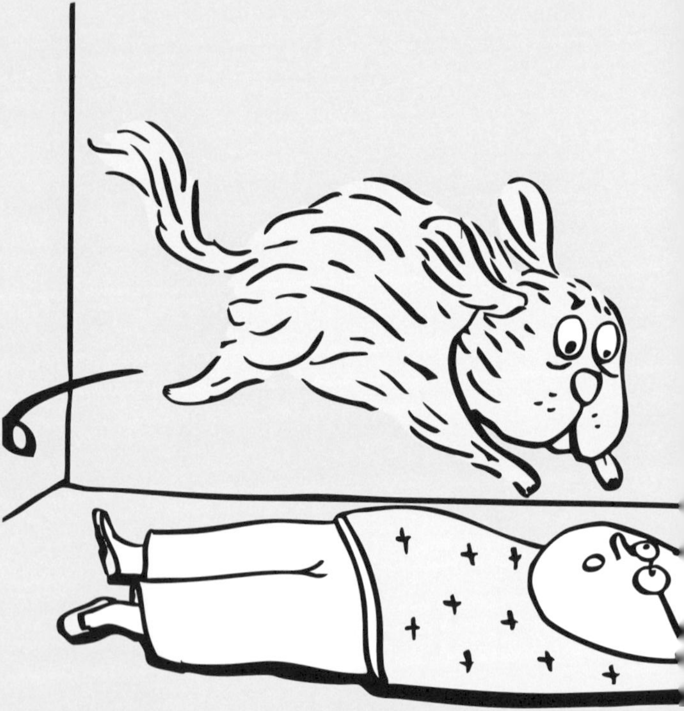

How can I persuade my dog that a simple handshake is enough?

If you were a dog and the doorbell started ringing, then new faces and voices appeared and a level of heightened excitement suddenly took grip on the house, I suspect you would bark and jump up on visitors, too. What is needed here is some basic training. If you spent 15 minutes daily working with your dog to stay on all four feet, even when excited, you could change their bad habits over weeks and months. The quick fix is to automatically put your dog away when visitors arrive. Once everyone has calmed down, then let them come in. They'll be much less likely to jump up in the same way. Once you have them trained not to jump up any more, you can begin to work on their handshake.

Why does my dog love chasing things?

Dogs are dogs and that means they love chasing. Give them a chance and they'll chase anything that moves, from cats to skateboarders; joggers to double-decker buses. It stems from their ancestry as hunting or herding animals.

Chasing comes with its own reward, if you think about it. The dog barks and chases something and it goes away...so in effect they've won! The fact that it was just passing by anyway is neither here nor there. If it's a cat they are chasing and the cat turns and has a go at their nose with its claws, it may stop them chasing cats. If cyclists or joggers are your concern, ask one or several of them to come by and surprise them by giving an equivalent biff on the nose with a blast from a water pistol or loudly shaking a plastic bottle with pebbles in it. This may put them off the idea in future before there is some kind of accident.

But it is you who needs to be in control when there are such concerns around. Get your dog to walk in a controlled way beside you, remaining focused on you at all times so that even cyclists whizzing by are not enough to distract them. Reward them with praise or treats when they have behaved well.

Chapter 4

A walk in the park

Just who's walking whom?

Try to take this on board straight away. Your dog is a dog, not a little human! They have been bred from generations of other dogs to work alongside humans and will be happiest when they are obedient. If you obey them, then every time they want to walk they will bark until they get their way...and so it goes with everything. It will drive you barking mad! So ignore them, never concede. You choose the time and place. You be 'in charge' on walks. If they are unruly, use a Halti type of head collar or a no-pull body harness to control them. Enrol for dog-training classes, which will teach you both how to deal with issues like Sit...Come...Stay... Walk-at-Heel. Once the two of you have mastered these essential tasks, you should repeat the exercise daily, putting them through their paces, reminding them of their training and rewarding them when they successfully complete a command. Remember, life is like a dogsled team. If you aren't the lead dog, the scenery will never change.

My dog often looks sad. How can I cheer them up?

Dogs are social creatures and are happiest when in contact with others, so, yes, the more time you spend with your dog the happier they will be. But that doesn't mean that when you are not with your dog they are unhappy. After a play or long walk with you, they will be perfectly happy to rest alone or groom themselves. Don't be fooled by their long or sad hangdog expressions. Many dog's facial expressions have been bred into them, so that Bassett Hounds or some Labradors will look sad even if you hand them a dish of cooked sausages!

Why doesn't my dog come back when I call them?

Place yourself in your dog's position. You are running free and having a great time and someone calls you back...you go and they put you on your lead and stop all your fun. What does it teach you? Probably that there is no incentive to come back when called next time! Always have a reward ready for your dog when they come back to you, and give them plenty of praise and attention. Do repeated recalls in training situations set up to be challenging, with other dogs and people around. Have food treats at hand to distract them or reward them even when they have learned to come back each time, so that they continueto associate coming back with a positive outcome.

They're such a lazy dog. What can I do?

Dogs are rarely lazy in the same way as a human, but if the exercise is boring, dogs are as likely as humans to refuse to do it. Find a way to make exercise fun. This could involve giving them regular treats during a walk on a lead, changing the route of your normal walk so that they have a new battery of smells to get excited over, or it could mean sessions of chasing a tennis ball, or just plenty of attention from you when you are out together. Find out what reward motivates your dog, and then include some of that reward for them in your daily exercise regime to encourage them to be more active.

How can I teach them that rolling in horse manure is simply not good?

Some people can't bear the smell of certain smells while others adore them. Some wear one scent to suggest one thing and on another occasion something subtler to communicate something else. It is just possible that the aroma of cow pats, fox pee or dead pheasant entrances your dog...so they roll in it, wishing to carry that scent with them.

Of course, it may hark back to ancient instincts when it was used as a way of carrying a message back to the pack that a certain animal is in the vicinity, or it may be used as a disguise when going to hunt the creature that deposited it.

Whatever your dog's reason for rolling in muck, the best tactic is to pre-empt it. Call them to heel or keep them on a lead until you have passed the problem. But from now on, go easy on the aftershave or perfume yourself. They may simply wish to smell as unpleasant to you as you do to them.

Where's the stop button?

Your energy will never match your dog's energy. A crafty dog walker will come armed with a nifty device designed to add to the dog's exercise with minimal exertion by the handler. An excellent example of such a device is a toughened tennis ball with a long-handled launcher that sends the ball flying into the far distance to be retrieved by your pet while you continue at your own pace. Vary the exercise with the different devices available from good pet stores, such as a Kong on a rope or a floating version, to encourage your dog to swim. Dog Frisbees also fly long distances, giving your dog the workout they require and you the relaxing stroll you deserve.

How can I stop them barking in the car?

Dogs feel very strongly that they should always go in the car with you, just in case the need should arise for them to bark in your ear! You probably brought it upon yourself by reacting to their *singing*, and they now think it's something you like. Some dogs get excited about travelling in the car; others get anxious. Ignore their noise and give them no reason to feel praised for doing it.

Begin their in-car training as a pup. If you have an adult dog that is excited in the car, you need to go back to basics. Start outside the car: training them to sit and stay. Keep them focused on you, as they sit still. Gradually move them towards the car, always rewarding them for being calm and relaxed. Put them in the car, then start the engine. The process should be done very slowly, not moving to the next step until the dog has succeeded in remaining relaxed.

Chapter 5

Dog gone!

Should I encourage my dog to be a bit more independent?

Dogs enjoy company, but they need to be encouraged to enjoy being on their own, too. If you have a dog that trails beside you, always under your feet, you need to take action, for the dog's sake. Make sure their bed is not right beside your bed at night. When you go for walks, use a long lead to encourage them to explore the world a bit further away from you. Share the dog care, so that they don't see you as *the only one.* Encourage them with rewards to settle down a small distance from yourself by placing objects between the two of you (such as chairs or big cushions). Create a secure, homey environment for them when you are not there. Leave the radio on and consider using a DAP diffuser beside their bed, sending soothing dog pheromones into the air around them. Lastly, reward them with a special treat before you leave, and soon you'll find that they are encouraging you out of the door.

My dog loves their own voice, but it is driving us mad!

When dogs get bored or want some human interaction, they bark, and barking has no volume control. To avoid a community meltdown, you need to do something about it now. Ensure that pooch has plenty of energy-depleting activities. Take them for a half-hour walk twice a day. Occupy their mind when they are outside by themselves with a frozen food-stuffed Kong, or a food-releasing puzzle cube dog toy, taking care to decrease the amount of food you give them at their normal mealtime to ensure that they have enough of an appetite to be interested in these food-rewarding toys. Failing this, try the citronella-spray collar, which is designed to squirt a lemon-scented spray at their nose every time they bark. Be sure to follow the instructions very carefully.

My garden is becoming as tunnel infested as the Great Escape!

One of the hardest things for an owner to do is stop their dog from doing something they really enjoy. Dogs love to dig holes. It is instinctive and harks back to their ancestor activities: making dens, looking for food or creating a cool spot in hot weather. Modern pet dogs dig for the simple reason that they enjoy it. Try to get them to learn where they can and cannot dig. Use a sand pit or soft earth bank where you can bury things for them to dig up, so they learn to dig on command and not through boredom, and reward them when they behave well. If you leave them alone in the garden, leave them with something to do instead of digging while you are away.

How can we make sure my dog has a good time when we're away?

Dogs enjoy holidays too! If you cannot take yours with you on yours, then let them have their own. There are plenty of great boarding kennels around and your dog can be pampered, hang out with the other dogs, be taken on walks, have one to one playtimes and all sorts of extras that won't cost an arm and a leg. Follow personal recommendations, then go and check the kennels out for yourself.

Look out for a high staff to dog ratio; is it friendly and clean; and do the dogs staying there at the time of your visit look happy and lively or sad and unkempt? Then, the only thing that need concern you when you return will be whether Pooch will want to go back to your boring old routine after having had such a good time at dog camp!

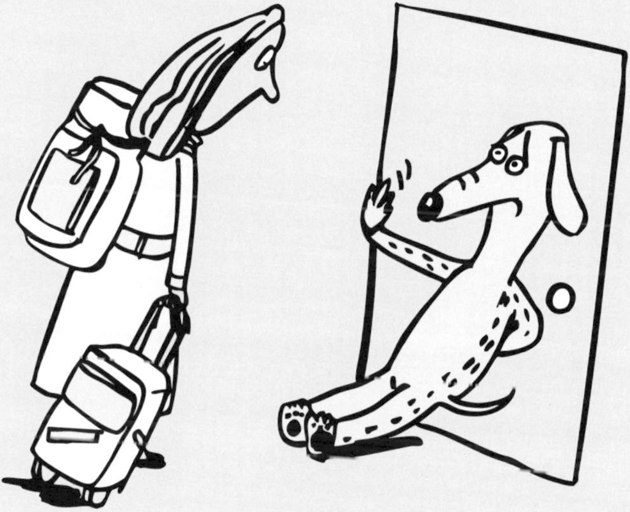

How can I stop my dog escaping from the garden?

There are two issues: hormones and confinement. If a female dog is in season, she is likely to try to escape to spend time with her suitors, while a male dog will travel many miles to sit waiting on her doorstep. Spaying and neutering are the obvious answers to the hormonal issue.

Some male (and female) dogs escape from the garden for no reason other than a sense of adventure. So if your dog is a bit of a tramp, you need to give them plenty of stimulation and exercise them for half an hour twice daily. A bored dog will get up to mischief.

Dog Day Afternoon... TV matinees for mutts?

Did you know that dogs see television much more clearly in the UK than in the USA? The more lines the TV signal is transmitted at, the clearer the picture for dogs. With High Definition TV, even more dogs are now joining in the pastime! This can, of course, be useful because some broadcasts can have a calming effect and hold your dog's attention while you get on with other things. You can even get your dog a show designed especially for dogs to watch and entertain them. These are easy to search out online. Just try not to let TV take over their life and slide into the introduction of TV meals! You could, of course, switch off the goggle box and get back to the good old days of taking your dog out for a walk and a game of fetch.

DR DOG

Chapter 6

Sick as a dog?

Should I introduce them to puppy love?

This is an area where it is wrong to put human thoughts and feelings into the minds of dogs. A male dog will have a strong urge to link up with a female dog when she is on heat, but when the scent of the female has gone, so has his physical urge. Dogs live in the moment and are unable to reflect on romantic matters. However, they are social animals and generally enjoy the company of other dogs. So, yes, it is a good idea to find a friend (girl or boy) for them. But if it's a girl, make sure both of them are neutered first.

How can I make my dog be more pleasant to their doctor?

OK, so your dog doesn't like going to the vet. Now let's try and view this from their perspective: You drag them along to a place they rarely go, they are forced to wait surrounded by all kinds of other animals, all equally bemused and nervous...some sick. Then they're called in to see a stranger who places them on a table. (Hang on...I'm not allowed on the table at home.) Then the stranger pokes them, inserts instruments into all kinds of orifices, and then they get a needle stuck in their neck.

It is a scary and unusual routine for a dog to go through. So the thing is to make it a better known and more pleasant experience for them. Drop by the vet from time to time if only to pick up some worming tablets or to get weighed so it is not all about 'the table'. Soon they and the staff will get to know each other and the place will take on a less threatening aspect, making it easier for everyone to relax...except about that place where the thermometer goes!

Walking when my dog hasn't had their puppy shots. Is it worth the risk?

We all want to show off our new puppy, but listen up: vaccinations protect puppies against a range of serious, potentially fatal infections, including parvovirus, distemper, infectious hepatitis and Leptospirosis. If a pup is exposed to one of these infectious organisms before they have completed their vaccine course, there is a significant risk that they will develop the illness. Ask your own vet when the vaccine protection will become fully protective. It's usually 7 to 14 days after the final injection, and from that point the world is their oyster.

Should I stop my dog licking my face?

Some people say that a dog licking a human is a submissive gesture, and that there is no harm in it. It has also been said that dogs carry fewer harmful bacteria in their mouths than humans, and that therefore the risk of infections being passed on may not be high. My view would be that an occasional accidental lick might not do any harm, but it doesn't seem sensible to make a habit of it, in either direction.

Dogs are scavengers by nature and pick up all sorts of foul-looking objects in their mouths when they are out and about. They are also avid cleaners of every part of their own bodies. I haven't yet seen a detailed analysis of the bacterial load on a dog's tongue, but I don't need to: it's common sense that some things are yucky, and having your face and lips licked by a dog is one of them.

How often should their bedding be changed?

It is said, 'Let sleeping dogs lie', but just where and what on is down to you. Most dogs, like teenagers, are happy to have their bedding cleaned by putting it through the washing machine no more than once a month. If, after this is done, the bed still has a nose-shrivelling pong, it may be linked to a bacterial or yeast infection of the dog's coat. You can control this with a twice-weekly washing of the dog using a special shampoo. So sniff around. If your dog's bedding and coat seem fustier than usual, pop him along to the vet; otherwise look forward to twelve straightforward dog-bed washdays per year.

Doggy breath?

Hey dog breath! There are three possible cause of halitosis in dogs:

1. Dental. Have a look at their teeth and gums. If you can see thick brown tartar on the teeth, and swollen, reddened gums, they need to visit the vet for a dental clean up.

2. Skin disease around the mouth. It can affect the folds of skin around the lower lips.You need to take them to the vet.

3. The digestive system. Foul-smelling gases are often produced while food is digested, and this often drifts up into the breath. If this is the cause, you should try a new type of diet.

To insure or not to insure?

Dogs can be clumsy and accident prone. They swallow objects that they shouldn't and chase cars, to name two of their hazardous activities. The resulting veterinary treatment can be eye-wateringly expensive. Invest in some pet insurance and peace of mind that if your pet subsequently has an accident or develops an illness, you will then be able to make decisions based entirely on what is best for them, rather than having to carry out an economic analysis of whether or not you can afford the procedure. There are many different policies out there, and you need to read the fine print carefully before signing up.

My dog always refuses to take their tablets. Any tips?

In both the animal and the human world some medications taste foul! There's no getting away from it. Thankfully there are some preparations for dogs that can be applied through the skin between the shoulder blades, but sooner or later something is going to crop up that requires them swallowing a tablet. This is why the hotdog was invented! Next time you have to administer a tablet, chop a cooked hotdog into several small pieces and secrete the tablet in one of them. Praise your dog and feed them two or three pieces in fairly rapid succession and then the piece containing the pill, letting them see you ready with another piece to follow it. Chances are they will gulp it down without even noticing.

Will my dog forgive me for having them neutered?

Dogs can survive without being neutured, but for the average household pet, it is the best decision. Neutering makes a dog's life simpler, and most dogs are better, healthier pets for it.

Mammary cancer, the most common cancer to affect female dogs, is prevented almost completely if they are spayed before their first season, and there are a number of illnesses affecting male dogs, including prostate problems and hernias, that virtually never occur in castrated animals. They can't thank you for sparing them from these, but then the concepts of blame, anger and forgiveness don't exist either in the bow-wow-now world of a dog. In short, they simply don't understand they are being neutered.

Talk about the subject in more detail with your vet, but it makes no sense to put your human thoughts into your dog's head.

About the author

Peter Wedderburn is a vet in a 'companion animal' practice, working only with pets. He has written for many newspapers, including the *Daily Telegraph*, and he regularly contributes to radio and television in the UK and Ireland. He lives in County Wicklow, Ireland, with his wife and two daughters and a menagerie of animals and birds.

His website is www.petethevet.com, and he's on Instagram @Pete.Wedderburn